This book belongs to:

_____

_____

Zodiac Signs:
A CAPRICORN STORY

(a Read-With-Ease book)

Written by B.M. Channing

Edited by A.M. Foster

Originally published as

"Zodiac Stories"

in 1899

by E.P. Dutton & Company.

Editor's note:

Title, names, spellings and some details may have been changed from the original version.

ISBN-13:   978-1979969727

ISBN-10:   1979969728

Copyrighted © 2017 by A.M. Foster, creator of the Read-With-Ease books. All Rights Reserved.

INTRODUCTION

The **Read-With-Ease** books are specially formatted for people who have one or more of the following conditions:

- Visual Impairments, aka Vision Impairments
- Dyslexia
- Computer Vision Syndrome (CVS), aka Digital Eye Strain.
- Struggling readers
- Tired, fatigued or strained eyes

For more **Read-With-Ease** books, visit:

http://readwitheasebooks.blogspot.com/

We also have a Facebook page:

Read With Ease Books

A CAPRICORN STORY

In the forest of North Rhine, there once lived a poor charcoal-burner with his wife and their only child, Hansel, a boy of ten.

There were no other children near enough for Hansel to play with but he did not mind this, for he made companions of the wild flowers and the wild creatures of the wood. He liked to watch the birds, the rabbits, and the frisky red squirrels, and he listened to the sounds they made, and imitated them so well that the little wood-folk answered him, and came to his call.

None of them feared him, because he never harmed anything. As he sat on a mossy bank, they would draw nearer and nearer until the bravest would drop down from a branch upon his head or shoulder, and the others, seeing it, would run up his arms, and perch here and there on his body.

Hansel's mother looked upon the strange power he possessed over the animals, as something very much like magic, and was afraid the fairies had a hand in it, and might carry the boy off to Elf-land someday. So she tied around his neck a pewter medal with the head of a saint on it, and taught him a prayer which he was to say to the saint if ever the fairies tried to bewitch him.

"I wish they would," said Hansel; "I would like to see the fairies."

"Never say that again, foolish child!" cried his mother. "Do you want to be carried off and turned into you don't know what?"

"I'm not afraid of them!" said

Hansel.

"Do as I tell you, just the same," answered Gretel Myers sharply.

"Why, yes, Mother, to be sure I will," he said, for he was an obedient boy.

One summer day Hansel sat on a bank under a favorite tree whittling a bit of wood into the shape of a goat. The live goat which he was copying munched the leaves of a vine close by. She was a pretty creature, and gave the milk which made Hansel's black bread taste so sweet at breakfast and supper.

He had finished his model, and was examining it with some satisfaction, when a stranger appeared beside him.

Hansel was startled, but not really alarmed. His first thought was of fairies, but he felt that fairies must be little and have wings, and would not wear suits of gray cloth, nor carry big white umbrellas, nor have such broad shoulders and long brown beards. Still, he was not very nervous, for he had not seen many people in his short life.

"Good-morning, my little man," said the stranger. "I have lost my way in your big tree-world. Can you tell me the road to Marl?"

"My father can," said Hansel shyly.

"And where is your father?"

"Yonder where the smoke is rising among the trees," replied the boy pointing. "He is a charcoal-burner."

"And you are a wood-carver, it seems," said the unknown man, stooping to pick up the wooden goat which Hansel had dropped. "Who taught you to carve, little one?"

"No one," answered Hansel,

blushing and looking down.

"A genius!" murmured the stranger to himself. "Will you take me to your father?" he added aloud.

The charcoal-burner rose from tending his fire, and took off his ragged cap as they approached.

He answered the gentleman's questions as to the way to reach Marl, telling him that the village was a good three miles away.

"So!" said the traveler. "I have walked far already and am tired. May I rest at your house before going farther?"

"With a good will," was Carl Myers's hearty response, and, striding along at a swift pace, he soon brought his guest to the little hut which served him for a home.

Gretel Myers was not upset to have a such a fine gentleman come in so unexpectedly. She set a stool for him at the roughly carved out table, placed a clean, coarse cloth upon it, and soon had a blue bowl of milk and a brown loaf of bread before him.

Herr Steiner praised the milk, and the pretty goat that gave it; and then he drew Hansel's carving from his pocket and asked if the child had really never had a teacher.

"No, indeed!" answered the parents in a breath. They were too poor to do more than clothe and feed their boy, and could give him no schooling.

"That is a pity," observed the Herr, "for he must be a clever boy."

Clever! Gretel was certain that no other boy like Hansel lived; and she ran to the cupboard and took out a number of his works of art, and stood them on the table proudly, for there might never be such another chance of showing off these wonders to someone who could appreciate them. There were little men and women, birds, rabbits, and squirrels, all, as the mother said, "The very living things themselves."

Herr Steiner looked at each with growing interest. And Hansel, hidden behind his mother, watched him in shy delight.

Presently the Herr looked up.

"Send your little lad out to play a while," he said, nodding towards the door. "I have something to say to you both."

Hansel went out, and walked beyond ear-shot, like the honest little fellow he was.

But he did not feel like playing. Instead he sat down under a tree, and wondered what was happening in the hut. He could not help thinking that they were talking about him. He began to feel frightened without knowing why. It seemed as if a long time passed.

Then his father called him, and he went slowly in, and saw that his mother's eyes were red.

"Get your Sunday suit on,

Hansel," said his father. "You are to go to the city with the Herr."

"Mother!" said the boy, looking at her with his pale face.

"Yes, dearest heart," she said, trying to smile. "The Herr will make a great man of you, he says. You shall learn to make beautiful statues like the saints in the church at Marl, and when you are rich, someday, you shall come and stay with us always."

Her voice broke at that, and Hansel ran into her arms, and Carl and Herr Steiner went outside and left them together.

In another half-hour, the artist and the boy set out together for Marl, Hansel carrying a little bundle in one hand.

There was a big lump in his throat, and the trees swam before his eyes, but he did not shed a tear until he cried himself to sleep that night in his room at the inn.

By and by came a letter from Dresden to the Myers family, giving good news of Hansel. Carl and Gretel walked to Marl to carry the letter to their kind old priest and hear him read it aloud, for they were unable to read it themselves.

The next letter was written by Hansel himself—which made them very proud.

"Ah! yes—but no one needs to be surprised at anything Hansel does," the mother declared, stroking the letter softly as it lay on her knee. "If ever a boy was cut out for greatness it is Hansel!"

The charcoal-burner only smiled at this, but he shared his wife's opinion.

The parents talked of little but their son and his prospects, and every night they went to rest with their hearts full of him.

But as the months passed into years, they began to long for a sight of him.

So the old priest wrote a letter for them to Herr Steiner, begging that Hansel might pay them a visit. And in time came a letter from Hansel announcing that he would be with them at Whitsuntide.

Then Gretel made great preparations, and there was nothing for a month but scrubbing and sweeping and mending.

"A poor place, Hansel will think

it, when all is done, after the fine houses of the city," sighed Gretel; but her husband shook his head.

"The lad is a good lad—he will be thinking how happy he was in the old days. Never fear but he will be glad to see his home."

And so he was. But the parents stared when he came springing in at the door one evening. Could this tall, well-dressed youth be their own Hansel—the very same Hansel? He was not long in convincing them, for his arms were about them, and his voice trembled as he called them by name.

After he had been kissed and embraced enough, and his height measured by his father's, he was allowed to seat himself on the roughly cut stool he used to sit on, and begin the story of his adventures in the five years since he left the forest.

His father sat opposite, silently smoking, but his mother drew her stool close beside his, that she might stroke his hand or the fine cloth of his sleeve.

Hansel had so much to tell that it was very late indeed before they retired to rest. He had seen so many wonderful things—such splendid galleries of pictures, and halls full of statues, and had met such great men; for the great men came to his master's studio to have their busts made; and, when some of them died, the city they had distinguished would commission the artist to make a life-size marble or bronze statue.

"And can't you make a statue?" inquired the charcoal-burner with a smile.

"Why not should the boy make a statue or anything else?" broke in Gretel sharply. "He can do anything with his hands. Did not the honorable

Herr praise the goat he carved on the very day he left us? See, Hansel—here it is, and a right pretty bit of work, too, I say—let someone else turn out a better!" So saying, the good soul ran to her corner dresser, and took down the wooden goat, and set it on the table before him.

"What a funny old thing!" the youth said, laughing, as he took it up and examined it. "Did I really do such bad work then? I will carve you a better goat now, if you will give me a bit of wood, little Mother."

And so he set to work straightway, and when he had

finished, the parents were obliged to confess that the new goat was finer than the old.

But Gretel's heart was a faithful one, and she kissed the clumsy old carving before she set it back on the shelf. She had taken pride in it so long, and it had comforted her when Hansel was far away.

"Old friends before new," was her motto.

When Hansel was starting to go back to the city, he slipped a purse of gold into his mother's work-worn hand.

"Eh, lad—what's this?" she cried, in dismay.

"It is all mine by good right, Mother; the Herr Meister gave it me for you. He says I have fairly earned it."

Gretel clung about his neck, crying. "My boy—my boy—! How can I lose you again!" she sobbed.

"Take heart, little Mother," he whispered, tenderly kissing her. "Someday when I am rich, and have a house of my own, you and father shall live with me, and we will all be happy together."

And then Hansel was gone.

He loved his parents, but his life grew more and more full and busy, and though he meant to go back and see them in another year, it passed, and the next, and the next, and he did not go.

He wrote, and his letters were the great pleasure of their life—the lonely, uneventful life of the great forest.

He told of the rapid progress he was making—of the orders for work which began to come in—of the praise he had received; of medals and prizes.

He traveled in other countries; his letters bore postmarks of Italy, France, England and America.

The good couple were much distressed that their son should go to such a far-off and dangerous place as America, which they had heard was a land of wild beasts and Indians. They were very glad when they heard of his return to Germany.

Eight years had now slipped away since they had seen him.

And in the summer of this eighth year, Hansel realized his dream of a house of his own—a house fit for a great artist, and one filled with his own beautiful handiwork.

The dining-room was paneled in wood carved by himself, the stair-rail was a great, twisting, twining wild-grape vine. But most charming of all was the stand for the tall hall-lamp. This was a magnificent goat, poised in the act of springing from a rock—the lamp between its graceful horns.

Yet Hansel had realized but one half of his dream. He had the house, but now he must bring his old parents into it; and one sunny morning he set forth with a glad heart for the great forest.

"Yes," he said to himself, "I have left it too long already. I must lose no more time. Life is short, and one cannot always have a father and a mother."

He reached the little village on the border of the forest late in the evening, and slept at the well remembered inn where, as a homesick child, he had cried himself to sleep so long ago.

Early on the following morning, he started alone and on foot for the charcoal-burner's hut.

An unaccountable sadness had come over him, and it deepened as he drew nearer to the old home.

At last he saw it. He took off his hat, and ran towards it, shouting—"Father! Mother!"

But no one answered. No eager faces appeared at the little dark windows.

He pressed on, he entered through the unlocked door. All was empty, silent, and deserted.

The picture of St. Hubert was gone from the wall; the plates and mugs from the dresser; the big, coarse linen-covered bedding from the low, rough bed.

A squirrel sprang down and out of the door, into the green world beyond.

Hansel's eyes filled with blinding tears. He flung himself on his knees by the cold hearth, whose dead gray ashes seemed a picture of the desolation of his heart, and laid his head on the old stool.

"God forgive me!" he cried. "God grant me the way to find them!"

Then he rose up, and retraced his steps to the village. The priest could tell him all.

The old priest was dead, said the hostess at the inn, but the Herr could see young Father Gottfried—yes, indeed.

So to the house of the Father went Hansel.

The young priest's face grew bright with interest.

The honorable Herr was the son of Carl and Gretel Myers! Oh so! He knew of the honorable Herr's great fame and renown. But where were Carl and Gretel Myers? Truly—had the Herr not seen them yet? They left to go to Dresden a month ago to see the honorable Herr. Was it possible that some mischance had befallen them?

It seemed too likely.

Hansel could learn little more, but guessing that Father Gottfried had been of much use to his parents, he left with him a sum for his poor which greatly astonished the worthy man as to leave him nearly speechless.

And now Hansel was back in Dresden, seeking his father and mother.

Three anxious days passed after his return, and then as he sat at dinner in his handsome dining-room, pretending to eat, but too sad to do more than play with what was placed before him, a servant brought word that a pair of beggars were outside, demanding to see the Herr, and refusing to be sent away.

"I have in vain told them that your honor is not to be disturbed. They insist that you open this package."

Hansel seized the little parcel, and tore it open with shaking fingers.

It was the old wooden goat!

"Father! Mother!" he cried, rushing to the door.

Yes,—there they were,—aged, weary, worn,—but the same loving souls as of yesteryears.

"And now we shall be happy together," murmured Gretel, her head on his breast. "The goat has done it!"

www.ingramcontent.com/pod-product-compliance
Lightning Source LLC
Chambersburg PA
CBHW031558210526
45464CB00003B/1329